How Does
Sunshine
Become
Electricity

How Does Sunshine Become Electricity

Junhao Chu
Chinese Academy of Sciences, China

Bo Hai
Shanghai Media Group, China

Chang Qin
Shanghai Media Group, China

Translated by: Zhongying Xue

Published by

World Scientific Publishing Co. Pte. Ltd.
5 Toh Tuck Link, Singapore 596224
USA office: 27 Warren Street, Suite 401-402, Hackensack, NJ 07601
UK office: 57 Shelton Street, Covent Garden, London WC2H 9HE

Library of Congress Control Number: 2022943355

British Library Cataloguing-in-Publication Data
A catalogue record for this book is available from the British Library.

Funded by B&R Book Program

《太阳能的光电之旅》
Originally published in Chinese by East China Normal University Press Ltd.

HOW DOES SUNSHINE BECOME ELECTRICITY

ISBN 978-981-124-685-2 (paperback)
ISBN 978-981-124-686-9 (ebook for institutions)
ISBN 978-981-124-687-6 (ebook for individuals)

For any available supplementary material, please visit
https://www.worldscientific.com/worldscibooks/10.1142/12543#t=suppl

Desk Editor: Nur Syarfeena Binte Mohd Fauzi

Let's explore the world of science and chat with great scientists, boys and girls, are you ready?

Academician Chu Junhao

Hai Bo

Qing Chang

We are students

Contents

/017

Try our best to use solar energy

In the past few decades, a myriad of research has been conducted to find ways to exploit solar energy. Solar cells have been fabricated to generate electricity, solar energy successfully converted into electricity. Here, I would like to invite you to survey which products utilized solar energy, and to conceive what the potential applications of solar energy will be in the future?

/063

The wonderful world of sunlight

The sun, irreplaceable to our earth, gives out light and heat. As we know, the spectrum of sunlight consists of red, orange, yellow, green, cyan, blue and purple. If you study it carefully, you'll find a lot of knowledge about sunlight.

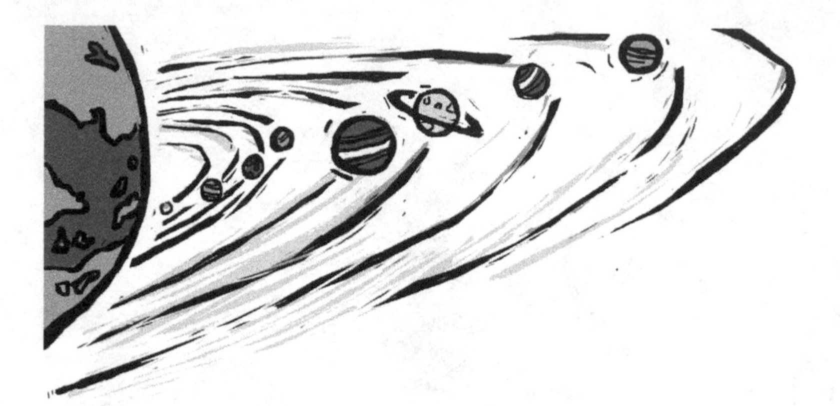

/085

The secret of photovoltaic conversion

When the electricity passes through the bulb, the bulb will glow. On the other hand, if a light beam shines on a semiconductor, the light may excite electrons and holes, i.e., generate electricity. Thus, light and electricity could convert mutually. Let's explore the secret of the conversion between light and electricity.

The explorer who grew up by Liwa River

Liwa River winding through East China Normal University (photographer: Cui Chu).

When I was a child, I always wanted to explore the unknown world. Back then, I was living in East China Normal University. A beautiful river, Liwa River, meanders through the campus, and I loved to play on the riverside. When I saw clouds drifting in the sky, I wondered what the outside world would be like. As a child, I had such a longing to explore something unknown, which had a significant impact on me later.

Hi, boys and girls. I am Hai Bo. We compile the series of 'Dialogues With Great Chinese Scientists' specially for you. In this series, we invite several great scientists in different research fields to share their stories and scientific knowledge. These scientists' stories are not only enthralling but also very inspiring. Maybe, in the future, some of you will become great scientists, too.

Qin Chang

Hai Bo

I am Qin Chang. Next to me is our distinguished guest — Academician Chu Junhao. It is my honour to introduce him to you. Today, Professor Chu will tell you his story. Does anybody know anything about Professor Chu?

Of course! Professor Chu is one of the chief editors for *Hundred Thousand Whys*. He is an expert on semiconductor infrared detection and solar energy applications.

Students

Professor Chu Junhao

Professor Chu is an academician of the Chinese Academy of Sciences (CAS), researcher at the Shanghai Institute of Technical Physics of CAS, as well as the Dean of the School of Information, East China Normal University. He specializes in the study of infra-red optoelectronic materials and devices. His research focuses on material physics and devices of narrow gap semiconductors HgCdTe and ferroelectric thin films, which have been applied to infra-red detectors. Besides, his research field also includes polarization materials and devices as well as solar cells.

Professor Chu is not only devoted to the original exploration in scientific research but also dedicated to popularizing scientific achievements to society and the public. He published nearly 100 popular science books and articles, *The Half of the Darkness — Infra-red* being one of them. Besides, he is the chief editor of *Energy and Environment*, one volume in *One Hundred Thousand Whys* Series (Sixth Edition), and the chief editor of *The Strategic Emerging Industries Popular Science Series.*

An explorer grew up by Liwa River

Qin Chang

Professor Chu, could you please share stories demonstrating your personal growth with us? As a great scientist, did you have some special or unique experiences in your childhood, which pushed you towards scientific research?

When I was a child, my family lived at East China Normal University. There is a river named Liwa River, crossing through the campus. At that time, the river was clean, and the environment was very beautiful. Since my parents were very busy, having no time for me, I often sneaked to the river to swim during Summer. Besides that, touring the vast campus with numerous buildings within was also my favourite. The whole campus was a paradise for us children. There was a building called the fifth dormitory, which had an attic. My friends and I would climb to the attic to chat and tell stories. Besides playing there in the daytime, we went there at night to see the sky. In those days, the city's lights were not so bright at night, and the moon and stars in the sky looked quite luminous. I found that some parts of the moon were dark and others bright. It looked like there were mountains and people on the moon. I felt curious about this kind of phenomenon.

When I was your age, a fifth-grader, I made a telescope by myself. I got a piece of cardboard and rolled it into a barrel, and then put a lens in either end. I used this shabby telescope to observe the moon and felt I could see the moon more clearly. I imagined about the shadow on the moon, picturing to myself a person living in the mountains.

I also read magazines, like *Science Pictorial*, and some biographies of scientists.

Student

You have such tremendous achievements. I want to know if you always achieved excellent scores when you were a child.

A cover of the Journal *Science Pictorial* (provided by periodical office of *Science Pictorial*).

Science Pictorial represents the oldest and the most influential comprehensive science periodical in China. It was first published in August 1933 by the Science Society of China for 'Saving China by Science' and has a history of more than 80 years. The journal has influenced generations of Chinese people.

When I was a student, I often went to the university library to read. Since there were not many people going to the library, a lot of dust piled up on the desks and books. Something must be done. So I thought of making a vacuum cleaner to get the dust off. Then, how do I make a vacuum cleaner? I was inspired by the working principle of the electric fan. We know that when we use a motor to propel the blades, the wind begins blowing, blowing the dust away. If the blades rotate in the opposite direction, is it possible to suck up the dust?

When we use an electric fan, the direction of the gas flow is determined by the rotation direction of the blade. If you are interested in that, you can do an experiment. Firstly, install a small fan to a motor. Secondly, connect a battery and the small motor by using wires, and observe the rotation direction of the blade. Then, switch the positive and negative poles of the battery and observe again.

Continuing my story. I made a very rough 'vacuum cleaner' with wood, batteries, motors and plastic sheets. I carried this vacuum cleaner to the library to test it. Unfortunately, this vacuum cleaner wouldn't work. My experiment failed, even so, I was still highly cheerful. After all, it is a suitable way to verify ideas through experiments. Here, I would like to give some suggestions to all the students. Please always be curious about the world because curiosity will inspire you to explore, take the initiative to read more books, and to create a lot of new ideas. If you have good ideas, I strongly recommend you practice and test your ideas.

Now, I still keep one of my report cards from primary school. My Science score was a 5 out of 5, with most other courses being 4. Since my handwriting was not very good, my calligraphy was a 3.

I didn't do very well in primary school. And when I first entered middle school, I was still quite naughty. Several years ago, when I was sorting out my documents, I found a report card. My report card had such a comment 'disrupting in class'. It was a nasty comment: I behaved very badly in class. My son asked me what was going on. I recalled such a story. One day, Li Xiahai, a very naughty classmate, asked me to shout with him in class, 'OK!', I said. Though there was no reason to do so, I still shouted loudly when he counted 1, 2, and 3, but he did not. Of course, I broke the regulations, so my teacher made me stand up and criticized me, and later left that comment on my report card. From grade two onwards, I was changing, especially from the second semester of grade two. In that year, I did a lot of maths exercises and became more and more interested in maths. I also read a lot of science books. Finally, I became a very hard-working student. So I aced all my courses from grade three and on, almost all 5 points, and kept this record throughout high school.

I began to learn physics when I was in middle school, and I was immediately attracted to this subject. I borrowed *Understanding the Universe through Modern Physics*, *The Eyes and the Sun*, and some other astronomical books from the library and devoured them. I started to read *Relativity ABC* and atomic physics books in high school. Although I could only understand a small part, I still read them with extreme fascination. So when I applied to go to college in 1962, I chose physics as my major. I hoped to go to Department of Physics of Fudan University, with my second and third choices being Department of Physics of East China Normal University and Department of Physics of Shanghai Normal College (Shanghai Normal University). However, my total score on the college entrance examination was not good enough. Although I got a full score in physics, the score in Chinese literature was relatively low, which dragged my total score down. Finally, I went to Department of Physics at Shanghai Normal College.

After graduation, I became a physics teacher in Shanghai Meilong Middle School and worked there for ten years. Over time, I did not give up studying physics, and I focused on theoretical physics. I organized a discussion group about elementary particles with some friends. Apart from discussing scientific problems, we also wrote popular science articles and books.

In 1978, China began to recruit postgraduates again, and I decided to register for the postgraduate school entrance examination. When I was in university, my major was four major mechanics and general physics. I chose semiconductor physics for my postgraduate research, though I hadn't learned anything about that before. So while preparing for the postgraduate school entrance examination, I began to teach myself about semiconductors. I took many notes and deduced the formulas in books by myself. Due to my solid foundation in university, I mastered new subjects very well and relatively fast. I remember my score in this subject was 90 points, the second-highest score of all the candidates. I was satisfied with that. I shared this story to let you know that you should learn the basic and main subjects well, which could facilitate your study of other related subjects.

There is no short-cut to success, and unremitting efforts are needed. No matter what you do, it's impossible to succeed easily. Let me tell you another story. When I was a postgraduate student, I once needed to design a semiconductor model. The premise of creating an excellent model is that every step should be conceived carefully, and every formula should be deduced. It was quite challenging for me at that time, but I did not lose heart. I put in a lot of time and energy, designing each part carefully, deducing what needed to be done with extreme patience, and eventually, all the principles were completely understood. So, it's pretty hard to get a good result. If you don't work hard, you can only do some easy, ordinary things. For scientific research, doing some innovative and distinctive experiments involves assiduous efforts. More often than not, the experiments don't produce the expected results, even resulting in total failures. Failure is not a monster. As long as we summarize the experience and try again and again, we can get the desired results.

Student

Do you have any special methods when you study?

For me, it is my belief to start one thing and stick it out. What we should overcome is the temptation to give up easily. If we resolve to do something, we need to set the right direction first. In addition, we must firmly believe that what we are doing is significant. Only in this way can we do it unremittingly and achieve good results.

Pupils from the Qingpu District wrote their questions on the 'question board'.

Boys and girls, I know you are actively thinking at your age. All kinds of questions crop up in your mind from time to time. They are fascinating and imaginative. That's so cool, as curiosity is essential for scientific research. Don't be inhibited by stereotypical results. Be brave and ask your question, that is, to think why before accepting it.

Of course, we should think and seek the result when we ask why. Rome was not built in a day. In the way of exploration, we should bear in mind three words: step by step. We should move forward, step by step, with basic knowledge and proper learning methods behind us. The famous scientist, Tu Youyou, spent decades studying and exploring. She and her team tested a lot of herbs, consulted a variety of books and materials, and finally they found Artemisinin and won the Nobel Prize. Success comes from conscious and conscientious efforts.

Try our best to use solar energy

In the past few decades, plenty of research has been carried out to find ways to use solar energy. People fabricate solar cells to generate electricity, and the conversion of solar energy to electricity is realized. Here, I would like to invite you to survey which products utilized solar energy and think about the potential applications of solar energy in the future?

Guide

Which products use solar energy? What are their advantages and disadvantages?

What is solar power generation? Which materials are used to make solar cells?

How much energy does a standard solar panel generate? What will Shanghai look like if all the power is supplied by solar panels?

Which areas of China are suitable for generating electricity by solar panels?

Before reading this chapter, you may think about the above questions and probably find the answers after reading. If you want to know more about these questions, please search online.

I learned from a book that solar cars have been built abroad. But it can only be driven in the daytime and not at night, because it cannot store enough electricity.

Student

Qin Chang

Is that true?

Solar energy — a new type of propelling energy

The Sun's importance to the Earth is self-evident. Being the nearest star to the Earth, the Sun provides us with light and heat. Without the Sun, there would be no life on the planet. Most of the creatures on the planet get energy from sunlight.

Why do we have to study about solar energy?

Mankind has long hoped to exploit energy from the Sun in a controllable, stable, and continuous way. As a new type of energy, we hope that solar power can replace coal to generate electricity and displace gasoline to propel cars, even aircraft.

We have made breakthroughs in the use of solar energy. Solar-powered automobiles and solar-powered aircraft have materialized. Recently, a solar-powered aircraft made a trip around the world without using a single drop of fuel and visited China on its way. Here is a news report about the solar-powered aircraft 'Solar Impulse 2'.

'Solar Impulse 2' completes its world trip.

At 4 a.m. local time, on July 26th, 2016, solar-powered aircraft 'Solar Impulse 2' completed an epic world trip and successfully landed at Al Bateen Airport in Abu Dhabi, the capital of the United Arab Emirates. 'Solar Impulse 2' is a long-haul aircraft that does not consume a single drop of fuel and solely relies on solar energy for continuous flight around the clock.

In March 2015, 'Solar Impulse 2' started from Abu Dhabi and began to fly around the world. It stopped in 16 cities in Oman, India, Burma, China, Japan, the United States, Spain, and Egypt, Chongqing and Nanjing. It eventually completed its flight of more than 40,000 kilometers and 500 hours around the world.

From Sohu News http://news.sohu.com/20160728/n461403046.shtml

From the news reports and an interview with the two pilots, Bertrand Piccard and Andre Borschberg, we knew that the wingspan of the air-plane is 72 meters long, only slightly shorter than that of the largest airliner Airbus A380, and the wings are covered with 17,248 solar panels. Due to its colossal size, it can only land at large airports. Enormous as it is, it weighs only 2.3 tons, roughly the weight of a family car, thanks to the use of new materials.

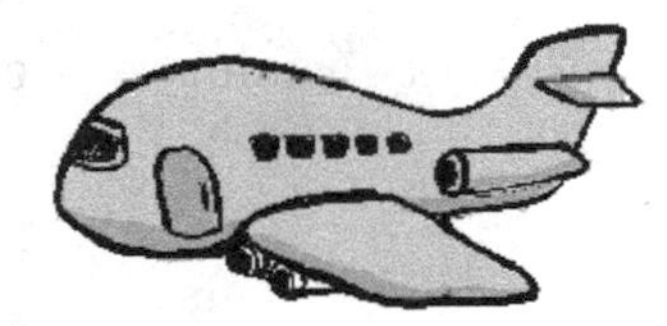

This aircraft is propelled by four electric motors, which are powered by solar panels in the daytime. Meanwhile, the extra energy generated by solar panels is stored in the lithium batteries, usable for flight at night. In this way, the solar aircraft flew day and night and

finally completed its global flight. This event plays a decisive exemplary role and shows the prospect of new energy.

Besides aircraft carriers, some cars also run on solar power. These cars are installed with solar panels and batteries. The electricity thus generated will be stored in batteries for driving the vehicle. In a sense, the solar car is a kind of electric vehicle. The difference is that the battery of the electric vehicle is charged by the industrial grid, and the solar panels charge the solar car. Some solar cars were displayed at Shanghai World Expo in 2010. Now there are solar air conditioners for vehicles, but they are not widely used yet.

Since solar powered aircraft can fly around the world, can we use solar energy for household appliances?

Student A

Student B

I notice many families have solar water heaters. I just don't understand how they work. Can we use this water heater on rainy days or cloudy days?

Try our best to use solar energy

From your questions, I know that you have observed life very carefully. The Sun contains huge amounts of energy. Scientists are making efforts to harness solar energy more efficiently and more conveniently.

Since the 18th Century, with the advance of modern physics, people have made significant progress in utilizing solar energy and made many unprecedented breakthroughs. Today, we have developed a variety of solar energy products. As long as we observe carefully and read books and watch the news, we can find many ways solar-power impacts our lives.

Conversion of light energy to thermal energy

In some rural areas of China, sunshine is abundant, so it is quite suitable for using solar energy. In recent years, many rural governments have promoted villages to use solar cookers, which represents a simple application of solar energy.

Yunnan Province is a sunny area. Even in Winter, the Sun is still dazzling in Kunming, the capital city of Yunnan Province. Solar water heaters of various sorts are installed on the rooftops in Kunming, indicating the popularity of solar energy utilization in this city.

Standing in the Sun, we feel warm, even hot, why? It is the thermal effect of sunlight. The irradiated object absorbs energy from light, and its temperature rises. When the Sun shines on the surface of the water, the surface temperature of the water also rises. That's the working mechanism of the solar water heater. As the sunlight is scattered and unstable, it is a challenge to gather heat from it and make our lives easier.

The process of utilizing heat converted from light involves collecting heat, storing heat and transferring heat.

We often find solar water heaters on some roofs. The solar water heater consists of several paralleled glass tubes, which serve as heat collection devices. This glass tube is usually made up of two layers of glass, with a vacuum in the middle. The inner glass tube has a coating, which can absorb the heat energy of the sunlight more effectively. The water in the glass tube will flow upward into the water tank after being heated, with the cool water from the water tank readily taking its place, which forms a heating circulation. During the daytime, the water in the solar water heater is heated and stored. When we turn on the tap connected to the water heater at night, hot water will flow out.

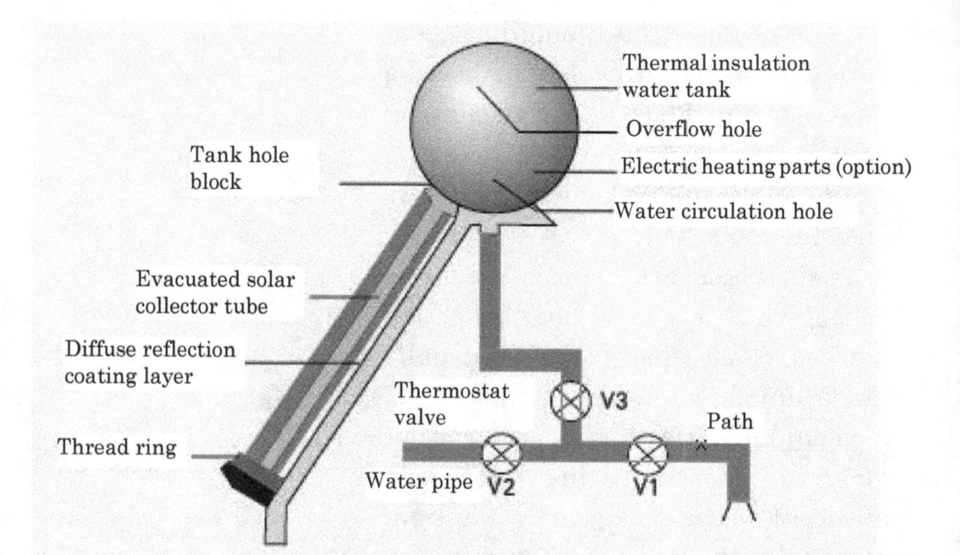

In theory, a solar water heater can be used in any area rich in solar energy. Still, in practice, this kind of water heater has not been popularized. Because of its intrinsic weakness — the solar water heater is at the mercy of the weather. The biggest drawback of solar water heaters is the difficulty in storing thermal energy and the inefficient utilization of heat energy. It can't be heated instantly. It will take quite a long time to heat the water, even when there is sufficient sunshine. In Winter and on rainy days, it becomes more difficult to use directly. To solve this problem, some solar water heaters resort to using electric heating technology to assist the process. When the water temperature cannot meet the requirements, electric heating is used to raise the water temperature. This seems to be a solution to the problem, but such an expensive water heater is not much different from the ordinary electric water heater in power consumption. Therefore, this kind of solar water heater is entirely unnecessary.

Photovoltaic conversion

Just now, someone asked whether household appliances could use solar energy. Theoretically, it is possible because electricity for household appliances can be supplied in various ways, including the public power grid and solar panels. If the electricity is generated by solar panels, that means we use solar energy.

This is a solar energy parking garage in Las Vegas, equipped with solar panels. Electric vehicles can be recharged here.

The solar cell, a device that converts light energy to electric energy, is usually made of semi-conductor materials. We call the process of converting light energy to electric energy photovoltaic power generation. The principle of photovoltaic power generation will be explained in detail in the following chapters. Here, I would like to share some news reports about the application of solar photo-voltaic conversion.

Urban residents use solar energy to generate electricity.

Here is a news report.

On March 28, 2013, Kong Qingbin, a citizen of Hefei City, started to run his 'family photovoltaic power station'. Kong Qingbin lives on the top floor of an 18-story building. So, he set up a photovoltaic power station on the roof of this building. This station includes seven solar panels in series, coupled with inverters and other equipment. According to news reports, on sunny days, besides meeting family requirements, his photovoltaic power station can supply power to a public power grid. On the other hand, on rainy or overcast days, he also needs to use the electricity from the public power grid.

On March 29, 2013, Zhongan Online, Journalist Zhang Wei and Wang Qian.
http://news.hexun.com/2013-0329/152632553.html

Photovoltaic power generation serves as a way to use such renewable resources as solar energy. People always hope for a new, clean and renewable resource, which has brought about a great improvement in photovoltaic power generation technology in recent years. As is reported in the news, many photovoltaic power generation enthusiasts design circuits, purchase equipment and install their own photovoltaic power stations. This is the origin of home photovoltaic power stations.

For a small-scale family photovoltaic power station, the design and installation are not complicated. Below is a schematic diagram of an independent household photovoltaic power station.

A schematic diagram of an independent household photovoltaic power station.

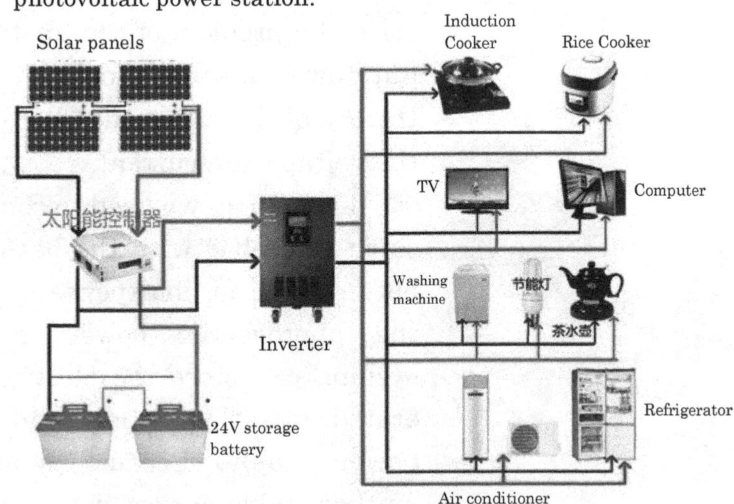

We connect the positive and negative poles of a solar panel by red and black wire, respectively. When the sun shines on the solar panel, the power output port will provide electricity.

So, can we connect electrical appliances directly to solar panels?

We seldom use electricity directly from solar panels because the current generated by solar panels is not stable. The output power of solar panels varies with the intensity of sunlight. Imagine that if a light bulb is directly connected to a solar panel, it will shine when beams of sunlight directly fall on the solar panel and turn dark when a cloud passes. Such unstable output power will undoubtedly affect the operation of electrical appliances.

With this being the case, how do we solve the problem of the unstable output power of solar panels? It requires the use of the solar controller to adjust the voltage and current.

In addition, we need to connect the solar controller with the battery. The battery, as an indispensable part of the photovoltaic power generation system, can store the electricity generated by solar panels and provide power stably. Because sunlight is neither stable nor consistent, we must find a way to store the generated electricity. By using a battery, the photovoltaic system could supply continuous power, even at night.

Here, we have another problem. We usually use alternating current (AC), and most household appliances are designed to work with alternating current. As solar panels supply direct current (DC), it cannot be used by appliances. To solve this problem, we install an inverter in photovoltaic power generation system to convert DC to AC. Then the AC from the inverter can be connected to the household appliances.

For a family photovoltaic power station, when the day is sunny, the electricity generated exceeds the family consumption, whereas it cannot meet the family's needs on cloudy or rainy days. How do we solve this problem? We need to connect the family photovoltaic power station to the public power grid.

In this scheme, we need a two-way electric meter. The electricity generated from the family photovoltaic power station can be transmitted to the power grid, or we can use the electricity from the public grid when the family photovoltaic power station cannot supply enough power, just like what we saw in the news report.

The household photovoltaic power station is only a small-scale photovoltaic power generation system. In China, there are many large-scale projects. In Gansu, Inner Mongolia, and Tibet, we have built large-scale ground photovoltaic power stations to transmit electricity to local residents.

The new shed in the countryside.

In rural areas, you may find a lot of sheds put up in the countryside. This type of shed is called a 'greenhouse'. When the Sun streams into the greenhouse, the temperature inside rises, much higher than that of the outside. Some greenhouses have been installed with a temperature and humidity monitoring system, which helps people better adjust the environment in the greenhouse. Many anti-season vegetables and fruits are planted in the greenhouses. For example, even in severe cold, tomatoes can be grown in greenhouses.

Nowadays, local governments are actively popularizing photovoltaic greenhouse or photovoltaic agriculture complementary greenhouse. So, what is photovoltaic agriculture complementary greenhouse? Compared with traditional greenhouses, a photovoltaic power generation system is installed on the roof of the new-type greenhouse. As different solar panels have varied transmittances, accordingly, other vegetables are planted in sheds. This greenhouse can not only produce fruits and vegetables but also generate electricity.

By laying solar panels on the roof, we can build a photovoltaic power station on a building. Considering the weight of solar panels, the roof of an ordinary building cannot sustain too many solar panels. To build a solar power station on a rooftop, the building should be designed and constructed carefully. Its roof should be made from a large number of steel structures, for example.

Shanghai Hongqiao Railway Station boasts such a rooftop, which used to be the largest rooftop solar power station in Asia and the largest single-body photovoltaic integrated project in the world. The photovoltaic power station is located on the platform-free canopy on both sides of the Hongqiao Railway Station. The solar panels take up 61,000 square meters, with the total capacity being 6.688 MW. Within a 25-year design life, the annual average power generation sent into the grid reaches 6,310,000 kilowatt hours, equivalent to saving 2,274 tons of standard coal every year, and reducing carbon dioxide emissions by 5,837 tons, sulfur dioxide emissions by 45 tons, nitrogen oxide emissions by 20 tons as well as dust emissions by 364 tons annually.

In recent years, photovoltaic power generation projects in China have developed leaps and bounds, so to speak. Not only Shanghai Hongqiao Railway Station, but also the stations of Hangzhou and other cities have built photovoltaic power stations. The photovoltaic integrated building project has also been popularized in urban communities.

In Shanghai, the media had such a news report. On July 18th, 2016, the first community-based photovoltaic power generation system was completed and

put into service on the rooftop of Chang Gao apartment building and connected to the public grid. Of course, this apartment is not unique. In Wuhan, Hefei, and many other cities in China, lots of buildings have a special design to use the solar energy better, save energy and reduce emissions. Reportedly, by 2017, China will have become the largest solar power producer in the world.

How much electricity can a solar panel generate in a day?

Student A

Student B

Is Shanghai suitable for photovoltaic power stations?

How much electricity can a solar panel generate in a day?

How much electricity can a solar panel generate in a day? This question is a little complicated, I will give some explanations below.

We usually use a square meter panel as a measurement standard to measure its power generation. In other words, we must first calculate the power generation per square meter of solar panels.

The material and installation method of solar panels are all important factors that affect its power generation. Each solar panel consists of many solar cells made from a variety of materials, the more common ones being crystalline silicon; such as polycrystalline silicon and monocrystalline silicon. Of course, lots of solar panels are made from something else, such as thin-film solar cells, organic solar cells, and so on. At present, photovoltaic power stations mostly use monocrystalline silicon solar panels and polycrystalline silicon solar panels.

Since the current and voltage of the monolithic solar cells are very small, they must be used in series or in parallel, i.e., connecting a number of solar cells to form a solar panel. The production of monolithic cells into solar panels requires a certain manufacturing process. Firstly, the solar cells are arranged in a framework (aluminum or other non-metallic materials) and connected with each other by connection components (such as diodes). Secondly, install a glass panel and back panel on both sides of the framework. Finally, fill nitrogen and seal the whole panel. We call this encapsulated panel a photovoltaic module or solar panel. The solar panel is the core part of the photovoltaic power generation system. Its photoelectric conversion efficiency (IPCE) represents one of the most important factors to determine its power generation. According to the relevant national standards, the IPCE of polycrystalline silicon solar panels and monocrystalline silicon solar panels should be no less than 15.5% and 16%, respectively.

The amount of power generated by solar panels is also affected by many other factors, such as the intensity of solar radiation, the latitude of the geographic location, ambient temperature, sunshine time, and installation angle.

Take the Yangtze River Basin in my country (near 29°N) as an example, the solar radiation energy of one square meter in autumn is about 1367 W, and the solar radiation intensity of the earth's surface may be less than 1000 W/m^2 after the atmospheric attenuation. If the IPCE of polycrystalline silicon solar panels is 15.5%, theoretically, one square meter of solar panels can generate electricity: $1000 \times 15.5\% = 155$ W. If the average daily sunshine time is 6 hours, it can generate $155 \times 6 = 930$ W h, roughly 1 kilowatt hour.

We can also say that in the same season, if this solar panel is placed in Qinghai and Tibet, it will generate about 1.5 kilowatt-hours of electricity per day. Of course, this data is only theoretical data. The actual temperature change, line loss, and the loss caused by the conversion efficiency of the controller (or inverter) will affect the final power.

The energy of the solar radiation is significant to all life on the Earth, but the Sun is extremely far away from the earth. How can we measure the energy that the Sun radiates to the earth? Scientists use the concept of 'solar constant' to describe the intensity of solar radiation above the earth's atmosphere. The value of 'solar constant' is generally 1367 w/m^2, which is a relatively stable constant.

Now, we know the efficiency of solar panels. According to this data, do you think Shanghai is suitable for photovoltaic power generation?

Qin Chang

Student A

Shanghai is suitable for that.

I don't think so. In recent years, sometimes smog has been quite serious in Shanghai. The sunlight will be blocked, so Shanghai is not suitable for solar energy.

Student B

Is Shanghai suitable for photovoltaic power generation?

To a certain extent, Shanghai is suitable for photovoltaic power generation. Because of the number of sunshine hours in Shanghai, many places, such as Hongqiao Railway Station, Pudong International Airport, and some residential buildings, have installed photovoltaic power generation system. But there are so many people in Shanghai, and the demand for electricity is tremendous. If all the electricity is provided by photovoltaic power generation systems, you can imagine how many solar panels we need. We do not have such a vast area for photovoltaic power generation systems.

In light of the limited land area of Shanghai, we cannot rely solely on the local photovoltaic power station. The electricity used in Shanghai is mainly from the public grid. Then who provides power to the public grid? It may be nuclear power plants, hydro-electric power stations, or photovoltaic power stations.

Knowledge expansion

Photochemical hydrogen production

In the BMW Museum in Munich, Germany, there is a silver-grey BMW car. It's extremely cool, catching many visitors' eye. This i8 BMW car is not for sale. What's special about it? It is a vehicle powered by a hydrogen fuel cell, and hydrogen fuel is known as the 'power of the future car'.

Then how do we produce hydrogen fuel cells? Hydrogen fuel is the ideal clean fuel for human beings. Hydrogen burns, producing water, which has little effect on the environment. Compared with other energy sources, hydrogen production is relatively difficult, and large-scale production is expensive. How can we prepare hydrogen fuel with a relatively cheap method? This is a problem that scientists all over the world are striving to solve.

Hydrogen fuel cell vehicle.

Judging from the research and experiments carried out in various countries around the world, the technology of using sunlight to catalyze the decomposition of water to produce hydrogen is one of the important ways to obtain hydrogen energy in the foreseeable future. Sunlight cannot directly decompose water. It must use a photocatalyst to absorb solar radiation energy through the photocatalyst, and effectively transmit it to the water molecules to decompose the water to achieve the purpose of hydrogen production. Therefore, the technical difficulty of solar photocatalytic decomposition of water to produce hydrogen is to find a photocatalyst. Fortunately, scientists all over the world have devoted great enthusiasm to research this field. International hydrogen energy research is also an important field of new energy research and application, just like solar energy research. The combination of the two research fields represents the direction of new research.

Regarding the relevant knowledge of using solar energy to produce hydrogen, because more students need to have a basic knowledge of chemistry, we will not expand on it here, and interested students can check relevant information by themselves.

Focusing on practice

In this chapter, we present numerous examples of using solar energy in our daily lives. I don't know if you have noticed but many solar energy products are commonplace within our lives. For example, some street lights are powered by solar panels. Maybe there's a solar power station near your home.

A device that uses solar energy to trap flies (image taken in Hefei).

Let's design a plan to survey the use of solar energy in your community. Write the results in the box below, or attach a picture in the box.

Survey

Use of solar energy nearby			
Solar energy products	Outlook	Where do we find it?	Work mode
Solar fan		Home	Photovoltaic conversion

Focusing on industry

Investigating China's photovoltaic industry

After reading the content of this chapter, the students may already understand that in order to use solar power, a complete industrial chain and supporting industrial technologies are needed. In China, a relatively complete and advanced photovoltaic industry has been formed. There are a large number of companies that can produce raw materials, batteries

Street light powered by a combination of solar and wind energy in Kunming.

and supporting components needed for industrial development, and provide corresponding production design, power grid assembly and service.

Students, after you finish this chapter, if you are still interested in exploring, you can go to the Internet to find relevant information and find out whether there is a photovoltaic power station in your city. If it is possible, I suggest you visit the station.

Is there a photovoltaic power station in my city?

For example: I live in Shanghai. The roof of Hongqiao Railway Station is installed with a photovoltaic power generation system.

Wonderful world of sunlight

The Sun, irreplaceable to our Earth, gives out light and heat. As we know, the spectrum of sunlight consists of red, orange, yellow, green, cyan, blue and purple. If you study further, you'll find sunlight is well worth researching.

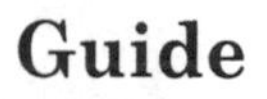

Guide

What is the essence of sunlight?

Why does the value of PM 2.5 become a major public concern?

Do you know why the sky is blue?

Professor Chu

Student A

Because there are a lot of ice crystals in the atmosphere.

Probably because the light emitted by the Sun is reflected.

Student B

Sky.

What is the essence of sunlight?

Just now, I asked a very common question. Why is the sky blue? Your explanations are wonderful, but it is better if we combine the two answers. Why do I ask this question? I want to know how the students solve a scientific problem.

As we know, the spectrum of visible sunlight consists of red, orange, yellow, green, cyan, blue and purple. Among the visible light, the wavelength of purple is the shortest. There are countless tiny particles in the air, which is like inevitable "impurities" in the air. The air scatters sunlight, and blue light is easily scattered to form a "blue sky".

When sunlight penetrates the atmosphere and reaches the surface of the Earth and our body, we can still feel the heat, right? It is the sun that brings us light and heat, i.e. solar energy. So what is the essence of solar energy?

The sun is essentially a thermonuclear reactor that continuously undergoes hydrogen-helium fusion. Electrons at extremely high temperatures continuously collide and vibrate with each other at extremely high speeds to generate electromagnetic waves. This electromagnetic wave is solar radiant energy, which radiates to space beyond the sun at the speed of light. What we call solar energy is solar radiant energy.

Dazzling sun.

Sun

For us, in addition to seeing light and feeling heat, scientists will also think about whether there are any other components in sunlight. Since there are seven colours in sunlight, then which colour causes the greatest thermal effects? Do you know?

I've heard of infra-red rays, but I don't know what infra-red is made of.

Student A

Professor Chu

Before answering your question, I want to ask whether anybody knows something about infra-red rays?

Strictly speaking, we cannot say what infra-red is made of, as infra-red is part of the spectrum of sunlight.

Student B

That's a perfect answer, I am so glad that some of you know so much about sunlight.

The sun radiates electromagnetic waves all the time, which consist of waves of different wavelengths. The wavelengths of some electromagnetic waves are quite short, such as X-rays used in X-ray examinations. Some electromagnetic waves are visible for human beings, 7 colours — red, orange, yellow, green, cyan, blue and purple — for instance. And the wavelengths of some electromagnetic waves are longer than those of red light, so they are called infra-red. The wavelength of electromagnetic waves can be as long as several meters, tens of meters, or even hundreds of meters. This is the case with WIFI. We can see the wavelengths of various electromagnetic waves from the electromagnetic spectrum chart below.

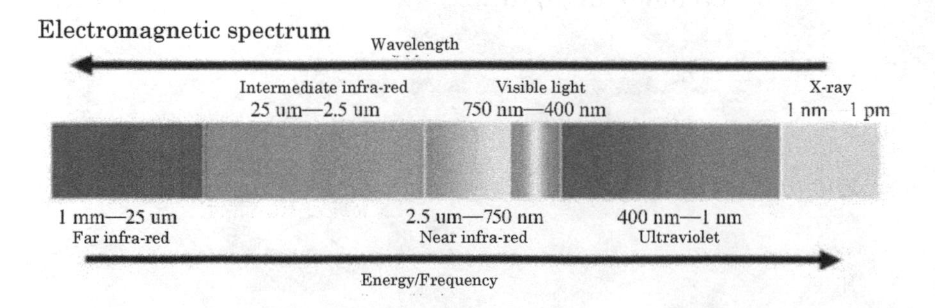

Therefore, infra-red is a kind of electromagnetic wave. Then what is the electromagnetic wave made of? It is composed of a constantly changing electric field and a constantly changing magnetic field.

I have another question, why are people so concerned about PM2.5?

Professor Chu

Student A

PM2.5 means the dust in the air.

It contains those pollutants in the air less than 2.5 microns in size.

Student B

Conversion of light energy to thermal energy

You are so great because all of you give wonderful answers. PM2.5 refers to particles in the air whose diameter is less than or equal to 2.5 microns. Why do I ask you this question? First, to remind you to protect the environment and form good habits and ideas of caring for the environment from an early age. Second, let you know that solar radiation will decay when it passes through the atmosphere. When the air pollution index is high, we will find that the sky is not so bright, and the sunlight is not so strong, which affects the utilization of solar energy.

Since the diameter of PM2.5 is extremely small, if people inhale PM2.5 particles for a long time, their lungs will be harmed.

Student

The utilization of solar energy mainly refers to the conversion and utilization of solar energy by modern industrial technology, including photothermal conversion, photovoltaic conversion and photochemical conversion. The first two are especially common and the technology is relatively mature.

In the previous chapter, we talked about heat collection. This serves as a way of using photothermal conversion. We adopt different lighting and heat collection designs to collect solar radiant energy and convert it into heat energy, which can be used to heat water, such as solar water heaters, and to heat air, such as solar greenhouses and solar air dryers.

Here, I want talk about solar thermal power generation projects. Why do we need to talk about the engineering? We often say that scientific research serves production technology. We have made many scientific breakthroughs in the laboratory, which is, of course, a very significant step forward. However, for our country, economic progress and national strength depend on the improvement of the productivity of the entire society. The improvement of productivity is inseparable from the support of engineering technology. There is still a long way to go from the results of the laboratory to the improvement of engineering technology in the entire industry.

The solar thermal power generation project is a very important position in the field of solar energy utilization. It is a large-scale industrial power generation method, that converts the solar radiation heat into electric energy and provides electricity for industrial production.

Large solar thermal power plants have been built and put into operation since the 1980s. Over the years, people have constantly refined technology to reduce costs and improve efficiency. Solar thermal power plants are distributed in the United States, Spain, Germany, France, the United Arab Emirates, India, Egypt, Morocco, Algeria, Australia and other countries. The Ivanpah Solar Power Plant in the United States may be the largest solar power plant in the world now, and we can read the related reports on this power plant.

A brief introduction to the largest solar power plant in the United States

'2014: Concentrated Solar Energy Annual Report' was published by the US Department of Energy. This report introduces five new types of concentrating solar power plants in the world, one of which is the Ivanpah Solar Power Plant.

Concentrated solar power plants

Ivanpah Solar Power Plant uses concentrating solar power technology. First, the sunlight is gathered from the reflectors and transmitted to the receiver, and the moving liquid in the receiver will be heated and turn into high temperature and high-pressure steam. Part of the steam is transferred to turbine generators, which drive the generator to rotate and generate electricity, and the rest is transferred to the heat accumulator, which stores thermal energy.

Ivanpah Solar Power Plant is located near Ivanpah, California, and began operation in February 2014. The power plant uses 300,000 computer-controlled mirrors to reflect sunlight on a boiler installed on top of a 139 meter high tower. The Ivanpah Solar Power Plant has an annual production of 3,920,000 megawatts, which is expected to be used by 100,000 American households.

Excerpted from http://solar.ofweek.com/2014-06/ART-260009-8490-28836591.html.

Discussion

Is solar energy a clean energy?

Solar energy is a renewable energy. Scientists have been studying how to make better use of solar energy. They hope that in industry, solar energy can be used to replace non-renewable energy such as oil, coal and natural gas.

So will the solar power plant have a small impact on the ecological environment because of the use of solar energy?

Here is a related report.

The world's largest solar power plant becomes a 'death trap'. Birds flying over are scorched.

The Ivanpah Solar Power Plant is the largest solar power plant in the world, located in the Mojave Desert on the border between California and Nevada. It covers an area of 8 square kilometers. Recently, the power plant has started operation, but the workers find that birds flying over the power plant will be burned. It is tested that the air temperature above the solar panels used for collecting sunlight was about 537°C. At present, dozens of birds and other wild animals have been found scorched. Environmentalists say there is ample evidence that the Ivanpah power plant can scorch birds flying in the area. It is reported that when the power plant is working, about 300,000 solar mirrors reflect sunlight and heat the water in the boiler at a high altitude to generate steam to drive the turbine to generate electricity.

Excerpting from Sohu News. http://mt.sohu.com/20170306/n482518318.shtml.

Do you have any comments on the news? We may have read a lot of information from popular science books and websites about the advantages of solar energy, and here, I want to ask you to consider there are any unfavourable factors for the ecological environment solar energy utilization. What do you think? Write down your thoughts in the box below.

The secret of photovoltaic conversion

After the current passes through the bulb, the bulb will glow. When light shines on the semiconductor, electrons and holes are generated in the semiconductor, and electricity is generated. There will be a conversion between light and electricity. Let us follow the photoelectric conversion to explore the secrets between them.

Guide

How does light convert into electricity?

How to store electricity?

Is it possible to use the photoelectric effect to create a perpetual power generator?

There is no air in the universe. How does the sun burn? Will it burn out?

Student A

I know that electricity can pass through a light bulb to make the light bulb glow, so how does light become electricity?

How can solar panels absorb sunlight and use it as an energy source?

Student B

How does light convert into electricity?

Solar cells are made of a kind of special material called semiconductor. Do you know semiconductor?

I know! A lithium battery is made of semiconductor material. I will not be shocked when I touch it, as it sometimes does not conduct electricity.

Student

There are two kinds of materials that are common to us. One is conductors, such as copper and iron. They can conduct electricity, i.e. the current flows through them easily.

Another is insulators, like leather or wood, which do not conduct electricity.

Apart from these two kinds of materials, there is another one, whose conductivity is between conductor and insulator, called semiconductor. The conductivity of a semiconductor varies with the conditions. Light does not shine on its surface, it does not conduct electricity. When the temperature rises, it will conduct electricity. Incorporated with certain impurities, it will also conduct electricity.

Photoelectric effect of Semiconductors

How is light energy converted into electric energy? You can consult some popular science books to learn about the remarkable discoveries of some great physicists and their epoch-making inventions. The emergence of solar cells and the development of photovoltaic industry are related to two major breakthroughs in physics.

Photovoltaic effect and semiconductor PN junction

In the 19th century, physicists made numerous significant breakthroughs in the study of electricity, the voltaic battery being one of them, which makes the current easier to store, and laid the foundation for the invention of dry batteries and accumulators. After the emergence of voltaic batteries, lots of physicists began experimenting with this invention. In 1839, French physicist A.E. Becquerel, while experimenting with a voltaic battery, found that when the Sun was irradiated, an extra electromotive force was generated between two pieces of metal immersed in the voltaic battery. He called this phenomenon the photovoltaic effect. The photovoltaic effect is the theoretical basis for the production of solar cells.

By the early 20th century, a large number of physicists were studying crystalline silicon materials. In the 1930s, physicists at Bell Labs made a series of great inventions in the process of studying this kind of material.

The first transistor was born at Bell Labs on December 23, 1947, and the three greatest inventors in the history of the semiconductor industry, William Shockley, John Bardeen and Walter Brattain won the Nobel Prize in Physics in 1956. They pioneered the path of the semiconductor industry, pushing the latest inventions from laboratory research to industrialization, and ultimately promoted the rapid development of science and technology in the whole of human society. All kinds of electronic products, including sensors, computers, communication systems, and so on benefited from several major inventions in semiconductor technology and the success of industrialization.

Shockley was later hailed as the 'Father of Semiconductors' by the media. In 1948, he published his thesis 'The PN Junction in Semiconductors and the Theory of PN Junction Transistors'. In 1950, his book *Electrons and Holes in Semiconductors* came out, which discussed the semiconductor PN junction in detail. PN junction is the core of numerous electronic components. All the semiconductor devices originate from PN junction.

PN junction — the heart of solar panels

Solar panels are made of semiconductor materials, which are a kind of power generation device that can convert light energy into electrical energy. There are many kinds of materials for manufacturing solar cells, the most common and widely used is crystalline silicon.

The earliest crystalline silicon solar cell was developed by the US Bell Labs in the early 1950s. Why was crystalline silicon adopted to make solar cells?

This starts with the physical and chemical properties of crystalline silicon. The chemical properties of pure crystalline silicon are relatively stable, but under the action of artificial technology, we can use doping to make pure crystalline silicon into N-type semiconductors rich in excess electrons, and P-type semiconductors rich in excess holes semiconductor. (For specific explanation, please read the book "The World of Chips: Exploring Integrated Circuits".)

What will happen if N-type semiconductors and P-type semiconductors form close contacts?

If N-type semiconductors and P-type semiconductors are connected tightly, a PN junction is formed, which is the heart of solar cells.

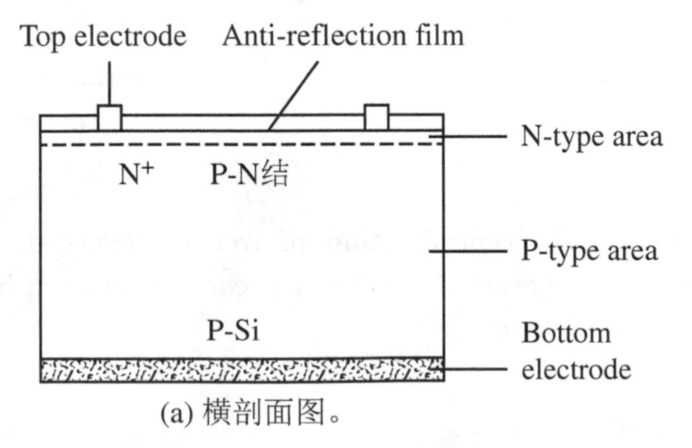

A schematic diagram of a typical crystalline silicon solar cell. The upper surface is N-type and the body area is P-type.

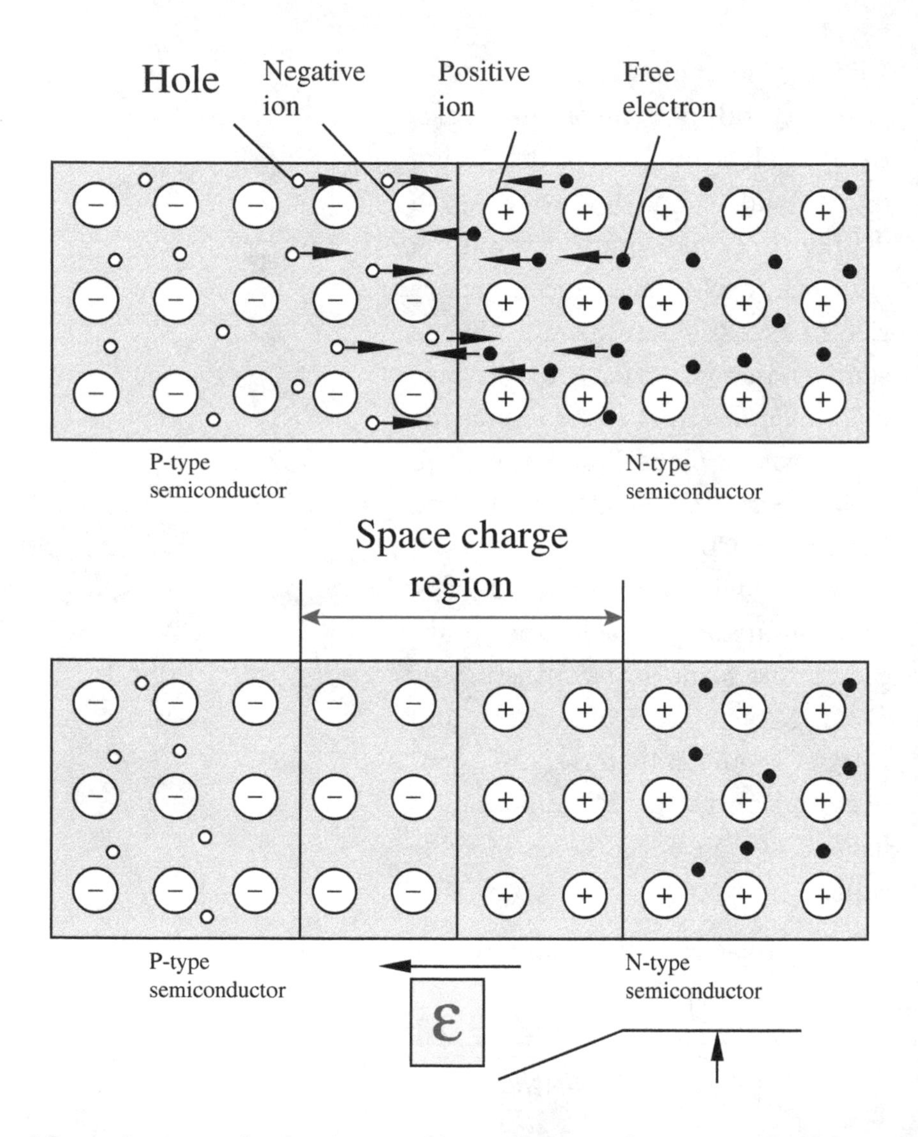

After doping, the concentration of free electrons in N-type semiconductors increases. So does the concentration of holes in P-type semiconductors.

When a PN junction is formed, electron and hole diffusion occurs near the interface of the semiconductor PN junction. As the superfluous electrons in the N-type semiconductor diffuse to the P-type semiconductor and recombine with holes, and the holes in the P zone diffuse to the N zone and recombine with electrons, a positive fixed charge zone and a negative fixed charge zone are left in the N-type semiconductor and P-type semiconductor, respectively, forming a space charge region (built-in electric field). Because of the existence of a built-in electric field, the diffusion and drift of carries will keep dynamic equilibrium.

Ions

All matters are made up of molecules, molecules are made up of atoms, and atoms are made up of nuclei and electrons rotating around them. When atoms or atomic clusters lose or acquire electrons, the charged particles called ions are formed. When electrons are obtained, the charged particles are named negative ions. When electrons are lost, the charged particles are called positive ions.

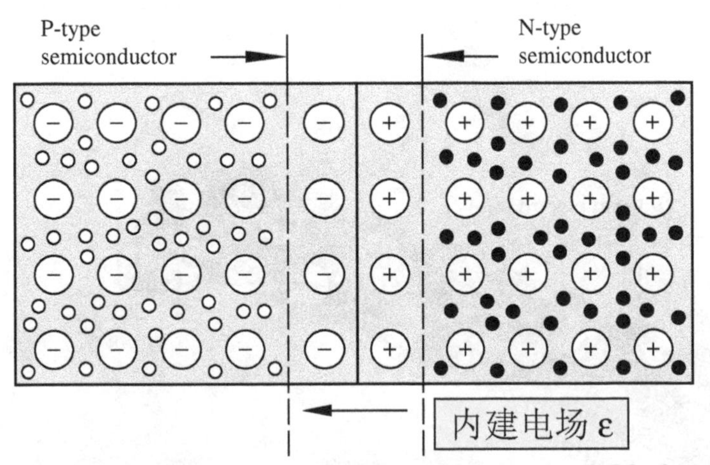

Because of the diffusion and drift, the electrons and holes in the PN junction will keep a dynamic equilibrium.

How does sunlight generate electricity?

We know that sunlight contains energy. When sunlight shines on the semiconductor materials, it can excite electrons and at the same time generate an electron-hole pair. In those semiconductor materials without PN junction, the electrons and holes generated by sunlight will recombine quickly and cannot form a current. On the contrary, if there is PN junction in the semiconductor material, the situation will be different.

When sunlight hits the PN junction zone of a semiconductor, the electrons it generates are negatively charged, and the holes generated are positively charged. The built-in electric field in the PN

PN junction of solar cell.

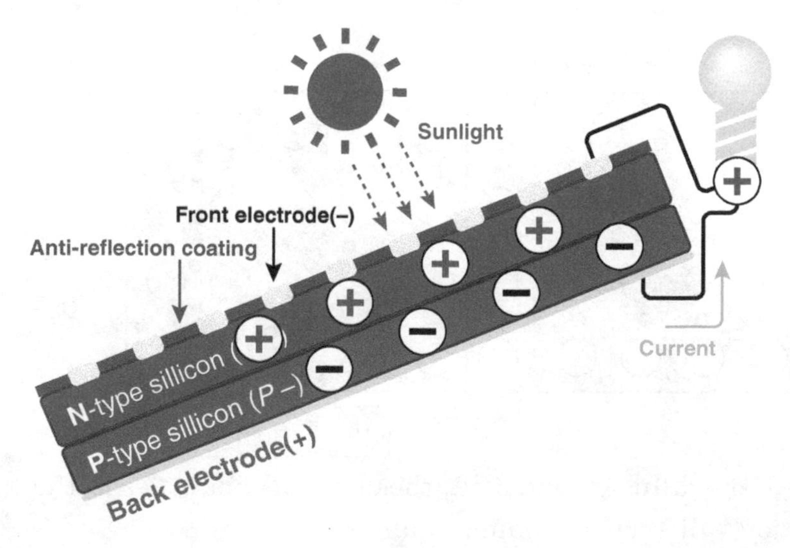

junction can pull the positively charged holes aside. The negatively charged electrons are pulled to the other side, then the two ends of the PN junction are just like our batteries. One positive electrode and one negative electrode, connected to the wire, can generate current. This is the principle of solar cell power generation.

How to store the electricity?

Has anybody ever played with a toy car driven by solar cells? There is a solar panel on top of that toy car. If the toy car is put in the sun for a while, it will run. But on rainy or overcast days, the car just stays put and won't run at all.

Therefore, if we want to apply solar power to industrial production, we should also solve the problem of solar power storage.

The storage battery serves as the most important device for storing electricity. In the history of physics, from the earliest voltaic battery to the later rechargeable iron nickel battery invented by Edison, the technology of the battery has been improving and successfully applied to the industry. For the photovoltaic power generation system that uses solar energy to generate electricity, the battery is an indispensable part. It has to be constantly charged and discharged to make the whole photovoltaic power generation system run.

Tesla Motors has invented a new lithium-ion battery that can be used in photovoltaic systems. With the rapid development of science and technology, especially the development of new energy technology, energy storage technologies that are compatible with them have also been introduced. Businesses are committed to improving battery capacity, performance and cost reduction. Tesla's lithium-ion battery is a direction for the development of energy storage technology.

Student

I want to know if I focus a flashlight on the solar panels at night, can I make solar panels generate electricity? If it works, is it possible that one day human beings can invent an energy-saving flashlight and shine light on the solar panels with it at night to achieve circular electricity generation?

Good idea. If we use a flashlight to shine on the solar panel to generate electricity, which later is converted into light, this system can circulate perpetually. It sounds very reasonable.

Hai Bo

Is it possible to create a perpetual power generator?

If we focus a flashlight on the solar panel at night, the solar panel can generate electricity.

During the process of converting light into electricity and further back into light, the energy will lose a lot. For example, there are many resistances in the circuit. When the current flows through the resistance, the resistance will heat up. As a consequence, part of the energy is lost as heat. Thus, this process cannot last forever.

There is no air in the universe. I want to know how the Sun burns? Will it burn out?

Student

Generally speaking, combustion is a chemical reaction. When we burn things, they react with oxygen and fire up, emitting light and heat. For instance, if we burn paper, we will see that the paper is on fire and becomes quite hot.

The burning of the sun is not an oxidation reaction, but a nuclear fusion reaction. There is a substance called hydrogen in the sun, and hydrogen has isotopes. One isotope is called deuterium and the other is called tritium. When these two isotopes have a fusion reaction, helium is formed and a small mass is lost. According to Einstein's mass energy relation formula, this part of lost mass produces enormous energy. When the fuel on the sun burns out one day, this fusion reaction will stop. Of course, other nuclear fusion reactions will continue. That is: helium nuclear fusion produces carbon, carbon continues its nuclear fusion to produce a series of atoms of silicon, oxygen and other atoms, and the last nuclear fusion is the formation of iron atom, which also means the doom of the sun. Since iron is the weakest and most stable element in nuclear energy, both iron fusion and nuclear fission are required to absorb energy instead of releasing energy, so when the sun burns to the stage of producing iron elements, the fusion reaction is going to stop. After that, small mass stars turn into dull white dwarfs and die, and massive stars end their lives in the big bang, leaving a neutron star or a black hole. Yet don't underestimate this last stage of the supernova explosion. Although it is an extremely short moment, it represents the most significant explosion in the universe. All the elements heavier than iron are produced at this moment, and there will be no life without the supernova explosion.

Focusing on related majors

Solar photovoltaic power generation project

In 1954, Bell Labs developed the first practical single crystalline silicon solar cell in the world, which started the era of the application of solar photovoltaic power generation technology. After a few decades of rapid development of photovoltaic power generation technology, the IPCE of solar cells has been greatly improved. Various countries have invested a lot of effort to develop the photovoltaic power generation industry, making solar photovoltaic power generation a large-scale industrial production. Services have formed a large-scale photovoltaic industry.

The photovoltaic industry in China has been developing quite rapidly. Here is a report from the Polaris Solar PV Network:

The title of 'The Largest Solar Power Station in the World' has changed hands several times in recent years. Now, China's 850 MW solar power station, the Longyangxia photovoltaic power station, is about to be completed and will come to the throne as the world's largest solar power station. The power station has covered 27 square kilometers of land and has been installed with nearly 4 million solar panels.

In 2016, with the photovoltaic grid-connected capacity adding up to over 77 GW, China, surpassing Germany, the United States and Japan, has become the largest photovoltaic grid-connected capacity country in the world.

From this report, we can feel the staggering progress of this industry, but the advance of any industry is inseparable from the support of basic science. Now let's look back on the development of this industry. From the discovery of the physical phenomenon of 'photovoltaic effect' to the invention of the semiconductor PN junction technology, and to the construction and production of the large-scale photovoltaic power station, in roughly 200 years, we have gone through a path of exploration and innovation.

Human beings have always been imaginative. Children today have the opportunity to reach out to the wider world and enrich their imagination. After acquiring more knowledge, you will be able to reflect on what you have imagined, knowing what is possible to materialize and what is out of the question.

When you grow up, you will need to further study professional knowledge systematically. You must know that from basic scientific research to final industrialization, this is a road that requires young people to struggle.

If you are interested in the content of this book, you can go online to search such majors as materials and electrical engineering in universities and to find out what you need to learn.

The majors of School of Information Science and Technology, East China Normal University.

A fantastic idea

Can we drive a car with sugar?

It is impossible to drive a car directly with sugar, but you can consider whether there is any indirect way to achieve it.

Professor Chu

Sugar is a kind of biomass. We can use biomass to generate electricity, or to extract a special oil. From this perspective, it is possible to drive cars with sugar.

The form of energy includes light, heat, electricity, magnetism, biological energy and so on, and they are inter-transformable. For example, light can be converted into electricity, electricity can be converted into light, light can also be converted into heat, and heat can also be converted into electricity. A solid command of scientific knowledge, the law of energy transformation in particular, makes it possible for energy to serve us better.

Classmates, is the movement of matter very interesting? Can you use your imagination to draw an interesting picture of the law of change in the form of matter movement that you think of?

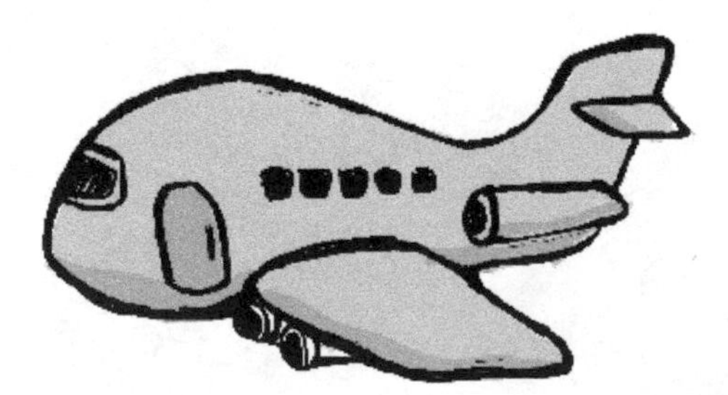

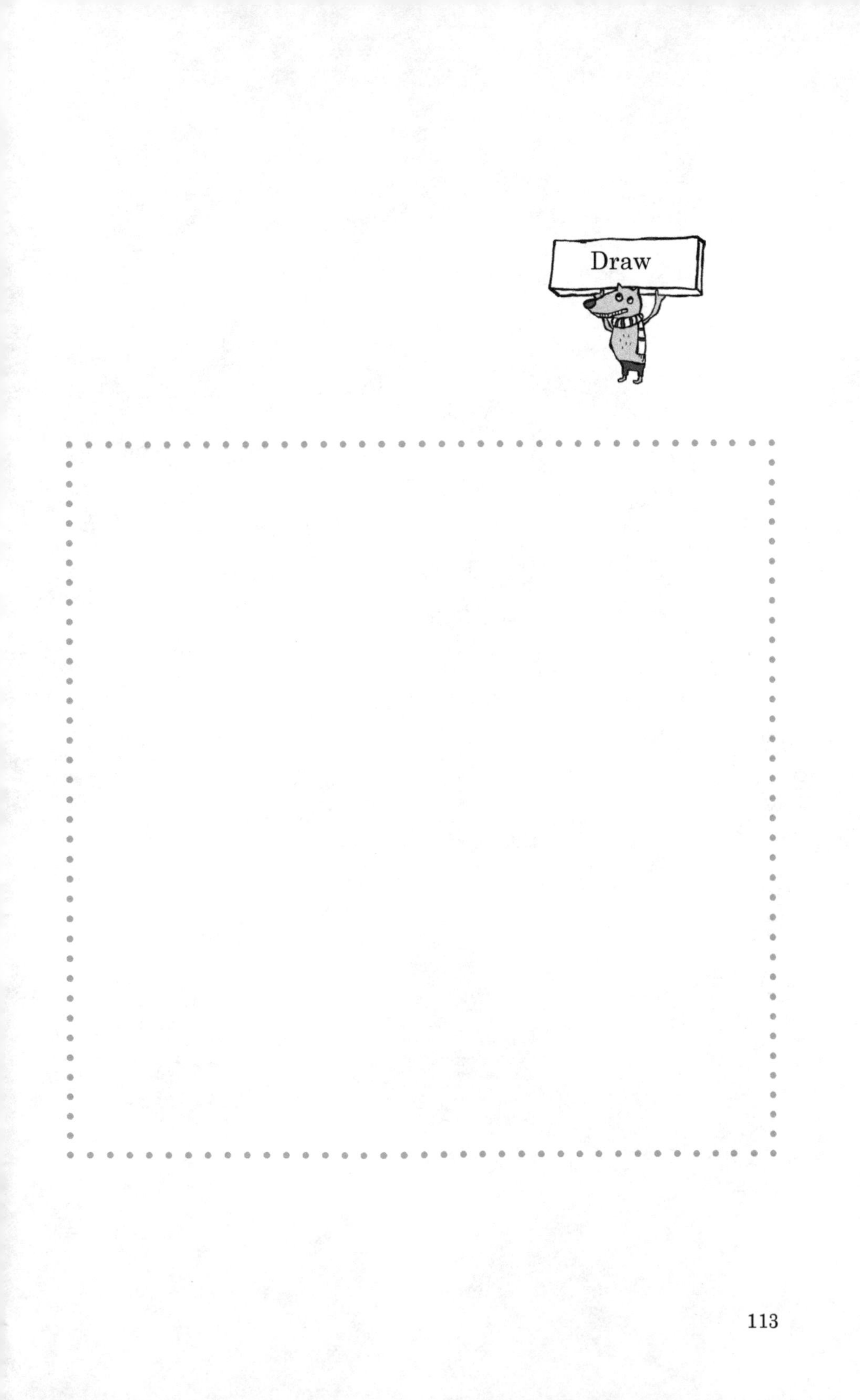
Draw